So You Want To Be A

SCIENTIST

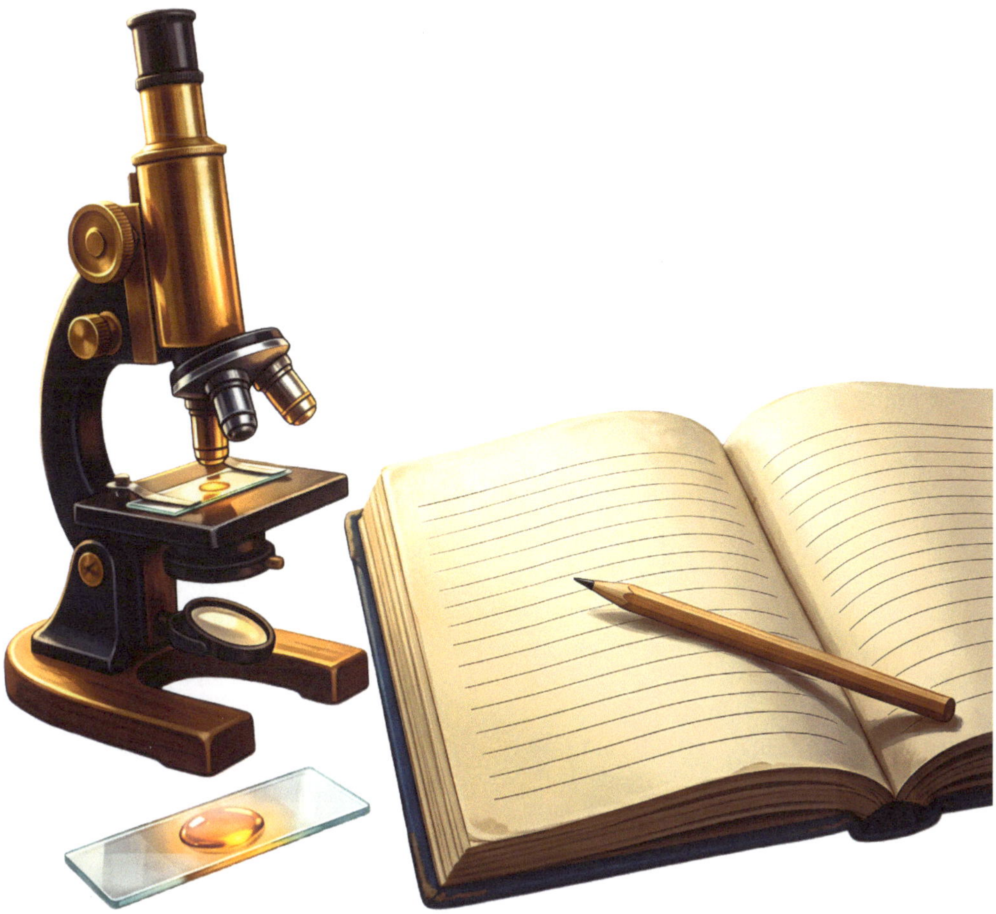

So You Want To Be . BOOKS

So You Want To Be A SCIENTIST

Linda Soules

For every kid who ever
turned over a rock
to see what lives underneath it.

ISBN 978-1-972766-21-7

What If You Were A Scientist?

So. You want to be a scientist. Good. Because you already are one.

If you have ever followed a line of ants to find out where they were going, you have done fieldwork. If you have ever mixed two things in a cup just to see what happens, you have run an experiment. If you have ever asked an adult "why?" three times in a row until they ran out of answers, you have done the hardest part of the job.

A scientist is someone who notices a small strange thing — a mold growing where it shouldn't, a chocolate bar melting in a pocket, a tree that looks lonely but isn't — and refuses to look away until they understand it. That's all. You don't need a lab coat. You don't need a degree yet. You need the question, the patience to follow it, and a notebook to write it all down.

Every question is a door. Science is the key. And behind every door is something nobody has ever seen before.

Why do birds fly in V formations? What makes ice slippery? Why do we dream? How does a seed know which way is up — even planted in total darkness?

Let's take the last one, which has a clear answer. Inside every seed are tiny grains of starch, heavier than the cells around them. They sink. The root grows toward where the starch settles; the stem grows away from it. A seed buried upside down in a flowerpot will still send its root down and its shoot up — because inside it, gravity is doing the measuring. Every plant you have ever seen figured out which way was up this way.

Scientists work across every kind of question: biology, chemistry, physics, geology, ecology, the brain. What they share is not a subject but a habit — the habit of following evidence wherever it leads, even somewhere unexpected.

Most scientists spend their careers contributing pieces to pictures much larger than any one person can complete.

What A Scientist Actually Does

A typical day in science is usually not dramatic. It is reading what other scientists have found. Designing experiments carefully. Running them, often many times, because single results aren't reliable. Analyzing data, checking for errors, consulting colleagues, and writing findings in papers others can evaluate and build on.

Some of the most important discoveries came from paying attention to what went wrong. Alexander Fleming left a petri dish out overnight; mold grew, the bacteria around it died, and he asked why. Result: penicillin, a medicine that has saved hundreds of millions of lives. Percy Spencer stood too close to a radar machine and the chocolate bar in his pocket melted; he asked why. Result: the microwave oven.

Here is the one most people don't know. In 1965, a chemist named Stephanie Kwolek was trying to make a new kind of strong plastic. The batch she pulled out of the machine that day looked wrong — thin and cloudy, like skim milk, where it should have been thick and clear. Everyone in the lab told her to throw it out. She asked to test it anyway. She spun the cloudy liquid into a fiber, and the fiber turned out to be five times stronger than steel. It was Kevlar. Today, police officers and soldiers wear it in bulletproof vests, and as a result, she has saved more lives than anyone can count. The work that accomplished this looked, on the morning she did it, like a failure.

The dramatic moment of discovery is the end of enormous careful work. And almost every time, the discovery was not the one the scientist set out to make.

In 1997, a forest scientist named Suzanne Simard did something strange in a British Columbia forest. She injected a harmless dye into the trunk of one fir tree and waited. A few days later, the dye showed up in a completely different tree, thirty feet away. Then another. Then another. Underground, connecting their roots, was a vast web of microscopic fungal threads — a hidden network through which the trees were sharing food, water, and even warning signals when insects attacked. Scientists call it the Wood Wide Web. Trees that appeared separate above ground turned out to be deeply connected below it — a discovery nobody made for centuries, because nobody dug.

What drives scientists is curiosity in its purest form — not the desire to appear knowledgeable, but the actual need to know. Science is what happens when that curiosity is given tools and methods and the patience to follow a question wherever it actually leads.

Where They Work

A bench scientist works in a research laboratory — a room with sturdy tables built to handle chemical spills, shelves lined with bottles of chemicals used in experiments, and special ventilated hoods that pull dangerous fumes away so the air stays safe to breathe. It has a particular clean smell that anyone who has been in one recognizes immediately. The bench is organized for the experiment in progress, each tool placed for the specific protocol being followed.

Field scientists work in the environments their questions require: remote forests, rock outcrops, mountain summits. What connects the pristine laboratory and the muddy field site is the organization around evidence. Everything is measured. Everything is recorded.

The Tools of the Trade

The Hypothesis

A scientist's most fundamental tool is the hypothesis — a specific, testable prediction about how something works. Before any equipment is used, the scientist determines the exact question being asked and the exact prediction being tested. This intellectual tool shapes everything that follows.

Laboratory Instruments

Microscopes magnify what's invisible to the naked eye. Pipettes measure tiny volumes with precision. Centrifuges spin samples to separate their components. Each instrument extends the scientist's senses into ranges the body cannot reach alone.

Data Recording and Analysis

Lab notebooks, data sheets, and computer databases are critical tools. The value of an experiment depends entirely on the accuracy of the records. Statistical software helps determine whether a result is genuine or random variation.

Peer Review

And peer review — having other scientists critically evaluate your work before it's published — is the quality control tool that keeps science honest. It's not hardware, but it might be the most important tool in the whole scientific enterprise.

The Best Parts of the Job

Scientists love the moment of discovery — the experience of being, for a brief time, the only person in the world who knows something. The experiment worked. The data shows the pattern. In that moment, before the paper is published, the scientist is alone with a new fact about the universe.

Science is also how we take care of each other. Every vaccine that saves a child's life started with a scientist asking a question. Every weather forecast that warns a family about a hurricane started with data collected year after year until the patterns became clear. Every glass of clean water, every safe bridge, every medicine in your cabinet — a scientist figured that out. Science shows us that the atoms in your body were made inside stars that exploded billions of years ago. That every human being on Earth shares 99.9% of the same DNA. Science is how we learn that we belong to something bigger than ourselves.

The community of science is something scientists describe with great warmth. Modern science is deeply collaborative; a single research paper can have thousands of authors. The paper announcing the discovery of the Higgs boson — a fundamental particle predicted by theory and confirmed by experiment — had 5,154 authors. It took three pages just to list their names. Every finding becomes part of the foundation on which the next finding is built.

The Hardest Parts of the Job

Most experiments fail. Not some. *Most*. A scientist might spend two years designing an experiment, months running it, and weeks analyzing the results — and the answer might be: nope. That's not it.

This is not a flaw in the system. This *is* the system. Science works because of failure. Every wrong answer eliminates one possibility and points you toward a better question. Every failed experiment teaches you something that no amount of reading or thinking could have taught you. Thomas Edison tested over 6,000 materials before he found the right filament for the lightbulb. He tried bamboo. He tried human hair. He tried beard hairs from his own assistants. He tried a thousand kinds of metal. On try 6,001, the filament glowed for 1,200 hours. Asked whether he was discouraged, he said he hadn't failed at all; he'd found six thousand materials that don't work. Learning to persist through extended periods of unrewarding work, to find the unfulfilling result scientifically useful even when it feels like nothing, is one of the most important and least discussed skills in research science.

Peer review is also essential to science, but can be difficult to experience. Reviewers finding significant problems require more work, sometimes months of it. This can result in the paper being rejected after months of waiting. The revision that addresses every concern may be returned with new concerns from a reviewer who seems to have read a different paper.

These experiences are familiar to almost every scientist who publishes, and none of them become less frustrating with repetition.

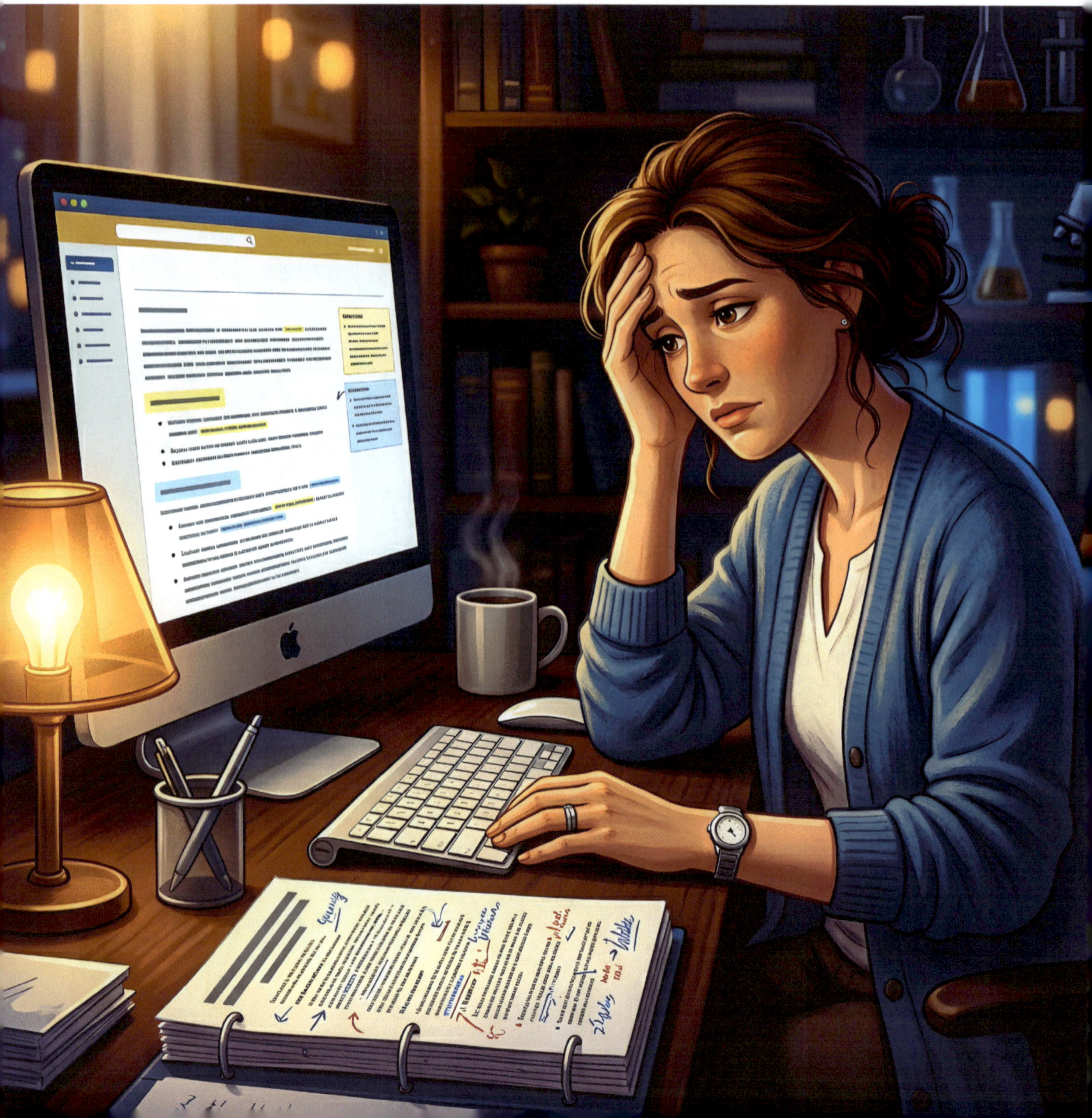

The Most Surprising Thing About This Job

he result a scientist publishes is almost never the experiment they set out to run. The riginal hypothesis is usually wrong in ways the experiment reveals and that redirect the esearch direction. The published paper describes a clean logical progression that gives no int of the many false starts and redirections.

eal science is messy — full of wrong turns, surprises, and lucky accidents. But when cientists write up what they found, they make it look like they followed a neat, straight ath from question to answer. The truth is, the path was never straight. Knowing that ifference matters, because it means the mess is normal — not a sign that something went rong.

nd the four most exciting words in science are not “I have found it.” They are: “That’s unny — what happened?” Every major discovery in the history of science began with a cientist noticing something that didn’t fit, refusing to dismiss it, and following the uestion wherever it led.

Here's something surprising: when other scientists try to redo experiments that have already been published, the results don't always come out the same way. In psychology, about six out of ten studies didn't fully work when someone else tried them again. In cancer research, fewer than half did.

This doesn't mean science is broken. It means that one study, by itself, isn't proof; it's a clue. Science gets things right over time, as more and more people test the same questions and the answers get clearer. One experiment is the start of an answer, not the answer itself. Science gets it right the way a crowd gets a right — slowly, with lots of arguing, and not all at once.

A Day in the Life

8:00 AM — Lab notebook updated first. Every day begins with writing: what happened yesterday, what it means, what today will test. The notebook is the permanent record that makes the science reproducible.

9:00 AM — Experiment setup. Variables controlled, equipment calibrated. Before the experiment runs, you've written down what you expect and why — because the experiment that confirms your expectation without your noticing teaches you nothing.

11:30 AM — Lab meeting. Every member of the research group presents progress. A researcher describes a result that doesn't fit her hypothesis. The group discusses it for forty minutes. The result that doesn't fit is the most interesting thing this week.

1:00 PM — Literature review: reading recently published papers. Science is a conversation that takes place in journals, and being part of it requires knowing what everyone else has said.

3:00 PM — Data analysis. One number in today's results looks very different from all the others. Instead of assuming it means something right away, you mark it and decide to run the experiment again tomorrow. One unusual number isn't an answer. It's a reason to look again.

5:00 PM — Writing: a section of the paper. You discover you can't describe what you found in a clear sentence, which means you don't fully understand it yet. This is useful to know.

The People You Work With

A research team — postdoctoral researchers, graduate students, lab technicians, undergraduate assistants — carries out the daily work together. A principal investigator leads the scientific direction. Scientists at other universities bring skills the team doesn't have.

Outside the lab, peer reviewers check the work before it's shared. Funders provide the money that makes the research possible. Science writers explain the discoveries to everyone else. And every scientist who ever lived and left behind what they learned is part of the foundation the next experiment stands on.

What makes a lab work isn't the equipment — it's what happens between the people. On Monday morning, someone shares a result that nobody expected, and the whole team spends an hour talking about what it might mean. A scientist in another country tries the same experiment and gets a different answer, and figuring out why takes six months of emails and new tests. A student asks a question the lead scientist has never considered, and suddenly the whole project changes direction. From the outside, science looks like one person having a big idea. From the inside, it feels more like a relay race in slow motion — each person carries the question a little further before passing it on, and the finish line keeps moving.

How You Help People

Science is the best method human beings have ever developed for figuring out what is actually true. Scientists build, piece by piece, the structure of understanding that everything else is built on. The medicine that cures. The engineering that holds. The technology that connects.

And there are questions that no one in the world can answer yet. What is consciousness? How does a brain — a three-pound lump of tissue — create the experience of being you? What is dark matter, the invisible substance that makes up 27% of the universe? How did non-living chemicals become the first living cell, roughly 3.8 billion years ago? Are we alone in the universe? Why do some kids get sick much more easily than others? Why does music give some people goosebumps and not others? Why does time feel like it speeds up when you're having fun? These questions don’t have clear answers yet. But they will.

The Most Important Qualities of a Scientist

Getting Excited When Things Go Wrong. This means meeting a failed experiment with "fascinating, why didn't that work?" rather than discouragement. The scientist who finds failure truly interesting eventually makes the discoveries that matter.

Telling the Truth Even When It's Inconvenient. This means reporting what you actually found rather than what you hoped to find — admitting what you don't know and not overselling what you do. All of science rests on scientists being honest, even when the truth is boring or wrong.

Writing It Down So Clearly Someone Else Can Do It. An experiment that can't be repeated by someone who reads your method is not science. Complete precision in describing what you did turns individual findings into shared knowledge.

Sharing Everything. This can involve publishing negative results, attending conferences, teaching the next generation, and building on each other's work openly. Science advances through sharing.

Working on Questions That Might Take Your Whole Life. Some questions take decades to answer. The scientist who finds real meaning in contributing one piece to a puzzle they may not live to see completed is doing the work at the depth it deserves.

Things You Might Not Expect To Need

Asking for Money (a Lot). Experiments cost real money. A microscope costs what a car costs. Running a team for a year costs what a house costs. So scientists spend a surprising chunk of their time writing to governments and foundations, explaining why their question matters. Most of the time the answer is no. They write another one. Then another. The scientist who can make her case in one clear page isn't doing paperwork; she's fighting for the right to do the science at all.

Writing. So Much Writing. A discovery that can't be communicated clearly to other scientists might as well not have happened. The scientist who writes clearly contributes to the shared project of human knowledge. The one who can't confines their discovery to their own lab.

Needing to Know *Why*, Not Just *What*. The question that begins every investigation isn't "what does the data show?" — it's "why is the world like this?" That deeper why is a philosophical question before it's a scientific one, and the scientist alive to it finds questions the purely technical mind never asks.

Famous Scientists in History

Marie Curie moved to Paris because women couldn't attend university in Poland. She worked in a converted shed with a leaking roof because the Sorbonne wouldn't give her proper laboratory space — and won two Nobel Prizes in two different fields, the only person in history to do so. She discovered two elements. She carried radioactive samples (releasing energy that can be dangerous to living things over time) in her pockets and stored them in her desk drawer, unaware of the danger. The research she produced with those samples saved more lives than it cost, many times over.

Carl Sagan could have spent his career speaking only to other scientists. Instead he spoke to everyone. His television series Cosmos was watched by 500 million people. He believed science was not separate from wonder; it was the deepest form of it. When the Voyager spacecraft was leaving the solar system in 1990, Sagan convinced NASA to turn its camera around and take a photograph of Earth from four billion miles away. In the picture, our entire planet is a single pale blue dot, smaller than a grain of dust. He wanted every scientist, and every child, to see how small and precious we really are.He made space feel like home.

u Youyou led a team searching ancient Chinese medical texts for clues to treating nalaria. She found a reference to sweet wormwood in a 1,600-year-old book, extracted ts active compound (artemisinin), and developed a treatment that has saved millions of ives. She tested the new treatment on herself before she would give it to anyone else. She von the Nobel Prize in Medicine in 2015.

Fun Facts

Newton invented calculus because he needed a mathematical tool that didn't exist yet. He invented an entire branch of mathematics as a side project.

Scientists at a lab called CERN speed up protons until they're moving at 99.9999991% of the speed of light — almost as fast as anything in the universe can go. At that speed, time actually slows down for the protons compared to clocks on Earth.

Marie Curie's notebooks from the 1890s are stored in lead-lined boxes and are too radioactive to handle without protective equipment. Researchers must sign liability waivers to view them.

If you uncoiled all the DNA from a single human cell, it would stretch about 6 feet long. Your body has 37 trillion cells. The total DNA would reach the Sun and back more than 60 times.

The scientific name for the common fruit fly — Drosophila melanogaster — means "dark-bellied dew-lover." This tiny insect has helped win six Nobel Prizes. Most of what we know about how genes build a body was discovered inside a creature that lives for thirty days and fits on a fingertip.

Scientists have revived bacteria found in 250-million-year-old salt crystals — microorganisms dormant since before the dinosaurs, brought back to life in a laboratory. Life is extraordinarily patient.

What You Can Do To Prepare Right Now

- **Start a science notebook**. Get a blank notebook and begin recording observations. What did you notice today? Write down one observation and one question every day. Date every entry. Congratulations — you're doing science.
- **Do experiments**. The kitchen is a laboratory. What happens when you freeze different liquids? How does temperature affect how fast sugar dissolves? Does a plant grow faster in sunlight or shade? Always write your prediction *before* you test it. Then compare.
- **Join citizen science projects.** You can count birds for the Cornell Lab of Ornithology (eBird), classify galaxies for scientists studying the universe (Galaxy Zoo), or track local wildlife through iNaturalist. Real science. Real data. Real impact — from your phone.
- **Ask questions and then look for the answers.** Don't just accept 'I don't know.' Go find out. The library, the internet, a teacher, a book — the answer is out there, and finding it is half the fun.
- **Fall in love with one subject**. Bugs. Rocks. Stars. Cells. Robots. Weather. Dinosaurs. Mushrooms. Whatever makes your brain fizz. Go deep. Read everything. Become the world's foremost ten-year-old expert. That depth is how scientists are made.
- **Read biographies of scientists** — not their discoveries, but their lives. You'll find that curiosity, stubbornness, and willingness to be wrong are the most important qualities. You can practice all of them right now.

Questions A Scientist Asks

Is this actually true —
or is it just what I expected to find?

What would prove me wrong —
and have I honestly looked for that?

What question has no one thought to ask yet —
and what would the answer change?

Could This Be Your Job?

When someone asks "Why do you want to know that?" — smile and say: "Because nobody knows yet. And I want to be the one who finds out." That's the scientist's answer. That's the scientist's instinct.

Do you love asking questions — especially the ones nobody knows the answers to? Are you the kind of person who needs to know why, not just what? Do you love the idea of letting the truth lead you wherever it goes? If so, ***then you might be a future scientist!***

Glossary of Scientist Words

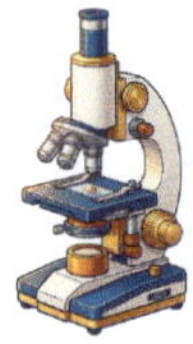

Citizen Science – Scientific research conducted with the participation of non-professional volunteers. Ordinary people contributing real data to real scientific projects. You can count birds, classify galaxies, or track species – and the data you collect actually matters.

Control Group – The part of an experiment that receives no treatment, used as a comparison against the group that does. Without a control, you cannot tell whether your result came from your idea or from something else entirely.

Data – The observations and measurements collected during an experiment. Without data, you don't have science – you have opinions. Data is the evidence that makes or breaks a hypothesis.

Experiment – A carefully designed test to find out whether a hypothesis is right or wrong. Change one thing, measure what happens, write everything down. Where science gets exciting.

Falsifiability – The requirement that a scientific claim must be testable in a way that could prove it wrong. If no possible evidence could ever contradict an idea, that idea is not science, however true it may feel.

Hypothesis – An educated, testable guess about why something happens. Not a random guess – a specific prediction based on what you already know, designed to be tested. Every great experiment starts with one.

Null Result – When an experiment finds no effect at all. Most experiments end this way, and a well-designed null result still teaches the world something real: this door is closed, look elsewhere.

Observation – Carefully noticing and recording something about the world. Curiosity and sharp observation are the two qualities every scientist needs most. They're also the two you can develop starting right now.

Peer Review – Other scientists checking your work before publication. It keeps science honest, rigorous, and reliable – and it can be difficult to experience, even when it makes the science better.

Replication – Repeating someone else's experiment to see whether the same result appears again. A finding that only ever happened once is not yet a finding. Science trusts what happens twice.

Scientific Method – The process scientists use to answer questions: observe, ask, hypothesize, test, analyze, and repeat. The most powerful thinking tool ever invented. Every discovery in human history was made using these steps.

Theory – In science, a well-tested explanation supported by enormous evidence. Scientific theories aren't guesses – they're the most powerful explanations science has.

A Note From the Author

Here is the harder thing this book has not quite said directly. Science is the most powerful way of understanding the world that humans have ever invented. But power is never only good, and the same knowledge that helps can hurt. Chemistry makes medicines. It also makes poisons. Physics built the electricity that lights your room. It also built weapons. Biology cures diseases. It has also, in some terrible chapters of history, been used by bad people to pretend that some kinds of humans matter more than others. Science itself doesn't come with a conscience. The conscience belongs to the scientist.

What I admire most about the scientists who inspire me is not how smart they were. It's how careful they were with what they knew. The best of them don't just ask "Can we do this?" They ask "Should we?" They tell the truth even when the truth is inconvenient for them. They share what they found, even when it would be easier — or more profitable — to keep it secret. They remember that the knowledge they uncover doesn't really belong to them. It belongs to everyone, including the people who come after.

So. Do you still want to be a scientist? Do you want to spend your life asking questions nobody has answered yet, and then finding the answers yourself? Do you want to be wrong a hundred times and keep going because you know the answer is out there, waiting? Then start now. Ask why. Write it down. Test it. Be wrong. Be wrong again. Be wrong in a new and interesting way. And then, one day, be right — and change the world.

The most important discoveries of the next century haven't been made yet. They're waiting for someone curious enough to find them. That someone might be you. Go find out.

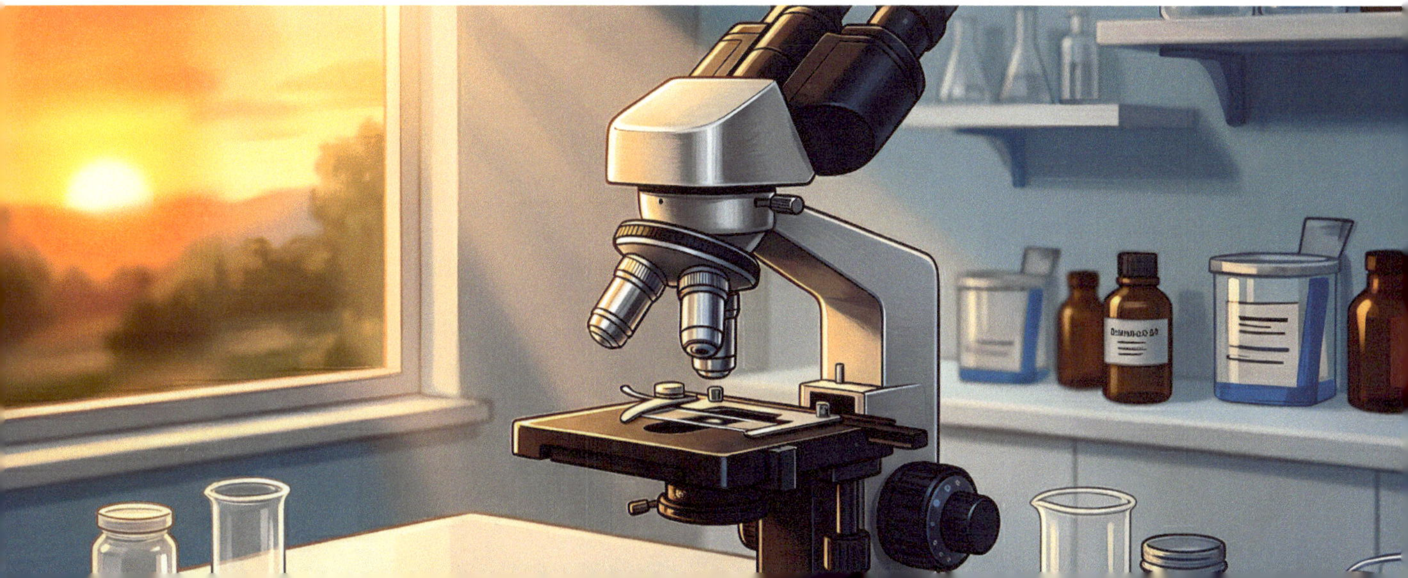

Want To Learn More?

Books, websites, and adventures for future scientists

BOOKS TO READ NEXT

The Girl Who Drew Butterflies: How Maria Merian's Art Changed Science by Joyce Sidman — In the 1600s everyone "knew" insects were born from mud. Maria Merian raised caterpillars and drew what she actually saw. Sibert Medal winner. Ages 10–12.

Hidden Figures: Young Readers' Edition by Margot Lee Shetterly — Four Black women mathematicians at NASA who calculated the trajectories that put astronauts in space. Science needs every mind. Ages 8–12.

Bomb: The Race to Build—and Steal—the World's Most Dangerous Weapon by Steve Sheinkin — Physics, spies, and the hardest question a scientist ever faces: we can build this, but should we? Newbery Honor and Sibert Medal. Ages 10–14.

The Scientists in the Field series (various authors) — Real scientists doing real work: tracking wolves, studying volcanoes, saving elephants. Fifty-plus titles; start with Sy Montgomery's The Octopus Scientists. Ages 9–14.

The Questioneers series by Andrea Beaty — Ada Twist, Rosie Revere, Iggy Peck: three kids who ask, build, and design. Start with Ada Twist, Scientist, and read it to someone younger than you. Ages 5–8.

WEBSITES TO EXPLORE

Science News Explores (snexplores.org) — Real science journalism written for readers 9 to 14, free and updated daily, covering every field from astronomy to zoology.

Science Buddies (sciencebuddies.org) — More than 1,200 free project ideas, searchable by subject, difficulty, time, and cost. Each one comes with a procedure you can actually follow.

Exploratorium Science Snacks (exploratorium.edu/snacks) — Over 300 hands-on experiments from San Francisco's Exploratorium, built from cheap household materials, each explaining exactly what's going on and why.

THINGS TO DO

Enter a science fair. Start with the one at your school. Society for Science keeps a directory of affiliated regional fairs, and placing well at one is how middle schoolers reach national competition. Pick a question you actually care about.

Talk to a working scientist. Skype a Scientist (skypeascientist.com) matches classrooms, libraries, and families with real researchers for free question-and-answer sessions. Come with your questions written down.

Go where the science is happening. University open houses, museum collection tours, astronomy club star parties, public lab days. Most are free, and almost all of them welcome a kid who asks good questions.

ORGANIZATIONS TO KNOW

Society for Science (societyforscience.org) — Runs the Thermo Fisher Scientific Junior Innovators Challenge for sixth through eighth graders, and the Regeneron Science Talent Search and Regeneron ISEF for high schoolers. Some past winners have gone on to win Nobel Prizes.

4-H STEM Programs (4-h.org) — Rocketry, robotics, environmental science, and more, for ages 5 to 18. Local clubs in every state, run through public universities.

SciStarter (scistarter.org) — A searchable catalog of thousands of real research projects that need volunteers. Filter by topic, location, and age to find one near you.

Every scientist was once a kid who asked "why?" and wouldn't accept "just because."

Keep asking.

SO YOU WANT TO BE A ...

196 Titles and Growing

Discover Every Book in the Series!

One for every dream, every passion, every curious mind. Which one will you read next?

◆ ◆ ◆

1 Astronaut
2 Veterinarian
3 Dinosaur Hunter
4 Video Game Designer
5 Firefighter
6 Zookeeper
7 Marine Biologist
8 Spy
9 Race Car Driver
10 Doctor
11 Robot Builder
12 Roller Coaster Designer
13 Pilot
14 Inventor
15 Magician
16 Shark Researcher
17 Police Officer
18 Chef
19 Explorer
20 Professional Athlete
21 Scientist
22 Surgeon
23 Actor
24 Stunt Performer
25 Toy Designer
26 Singer
27 Voice Actor
28 Deep Sea Diver
29 Vaccine Developer
30 Treasure Hunter
31 Crime Scene Investigator
32 Animator
33 Ballet Dancer
34 Astronomer
35 Waterslide Tester
36 Coder
37 Diplomat
38 Ice Cream Flavor Inventor
39 Fashion Designer
40 Candy Scientist
41 Animal Rescue Worker
42 Activist
43 Lawyer
44 Journalist
45 Psychotherapist
46 Psychologist
47 Architect
48 Teacher
49 Librarian
50 Geologist
51 Documentary Filmmaker
52 International Aid Worker
53 Public Defender
54 Social Worker
55 Judge
56 Recreation Therapist
57 Mediator
58 Child Life Specialist
59 Elder Companion Caregiver
60 Dog Trainer
61 Emergency Dispatcher
62 Botanist
63 Yoga Instructor
64 Influencer
65 Engineer
66 Board Game Designer
67 Horse Trainer
68 Code Breaker
69 Writer
70 Martial Artist
71 Chaplain
72 Basketball Player
73 Soccer Player
74 Movie Director
75 Esports Athlete
76 Rocket Scientist
77 Rock Star
78 Paramedic
79 Photographer
80 Nurse
81 Stop-Motion Animator
82 Artist
83 Meteorologist
84 VR World Builder
85 Mountain Climber
86 Flavor Tester
87 Foley Artist
88 Comedian
89 Animal Translator
90 Chocolatier
91 Dream Researcher
92 Baseball Player
93 Epidemiologist
94 Neuroscientist
95 Parkour Athlete
96 Choreographer
97 Mathematician
98 Animal Actor Trainer
99 Mayor
100 Hacker
101 Medical Researcher
102 Pediatrician
103 Coach
104 CEO
105 Salesperson
106 Event Planner
107 Economist
108 PR Specialist
109 Model
110 Wildlife Photographer
111 Smokejumper
112 Adventure Guide
113 Carpenter
114 Baker
115 Dentist
116 Ship Captain
117 Farmer
118 Train Engineer
119 Undercover Agent
120 Olympic Champion
121 Food Truck Owner
122 Tennis Player
123 Swimmer
124 Business Owner
125 Astrobiologist
126 Geneticist
127 Biochemist
128 Art Therapist
129 Radiologist
130 Speech Therapist
131 Occupational Therapist
132 Fitness Trainer
133 Sports Psychologist
134 Referee
135 Sports Scientist
136 Sports Agent
137 Financial Advisor
138 Real Estate Agent
139 HR Manager
140 Management Consultant
141 E-Commerce Specialist
142 Storm Chaser
143 Gymnast
144 Park Ranger
145 Figure Skater
146 Helicopter Pilot
147 Soldier
148 Wilderness Survival Expert
149 Whitewater Rafting Guide
150 AI Trainer
151 Playground Designer
152 Puzzle Maker
153 Beekeeper
154 Mechanic
155 Mural Artist
156 Lost City Finder
157 Stadium Announcer
158 Graphic Designer
159 News Anchor
160 Radio Host
161 Construction Worker
162 Electrician
163 Pottery Maker
164 Hot Sauce Creator
165 Food Stylist
166 Truck Driver
167 Sailing Racer
168 Fencing Champion
169 Pharmacist
170 Physical Therapist
171 Illustrator
172 Translator
173 Floral Designer
174 School Principal
175 Renewable Energy Engineer
176 Volcanologist
177 Beautician
178 Nutritionist
179 Interior Designer
180 Cinematographer
181 Editor
182 Auditor
183 Paralegal
184 Flight Attendant
185 Blacksmith
186 Fortune Cookie Writer
187 Crossword Puzzle Constructor
188 Cave Explorer
189 Drone Pilot
190 Search & Rescue Worker
191 Trapeze Artist
192 Color Namer
193 Instrument Builder
194 Master Builder
195 Professional Chess Player
196 Jewelry Maker

ABOUT THIS SERIES

What would I want to hand a curious child?

When I set out to create the *So You Want To Be A*... series, that single question was my North Star. I wanted to write something closer to a conversation than a career guide — a thoughtful, beautifully illustrated invitation into a world a child might one day step into. My aim was to create books that take young readers seriously and trust them with the real texture of a life's work: the wonder, yes, but also the quiet parts, the hard parts, and the moments of surprise that make every calling worth pursuing. Children are drawn to what is real. They can tell when a book is talking down to them, and they can tell when it is meeting them eye to eye. My aim, in every title, is to meet them eye to eye.

ON AGE RANGE

The series is officially designated for ages 10 to 14, but in practice it reaches well beyond that range by design. For younger readers ages 8 and 9, these books often become treasured visual companions. A child this age may not read a full title cover to cover alone, but they will linger over the pictures and absorb the parts that spark something in them, and shared reading with a parent or teacher opens the door to wonderful conversations about careers, ideas, and the endless question of what is possible. For middle readers ages 10 to 12, these books tend to land in their sweet spot, as children this age often read independently, dive into the details, and begin imagining themselves in the careers described. For older readers ages 13 and beyond, the series becomes a resource for real career exploration. And for the adults in the room, each book offers something unexpected, from fascinating facts to surprising history to memorable insights worth discussing as a family. In short, these are books that children can grow into rather than out of.

ON CRAFT

Every word in this series is written by me, and the research behind each title is extensive, drawing on firsthand accounts, current reporting, historical sources, and careful fact-checking. The illustrations are AI-assisted, and I want to be transparent about that because I know it matters to many readers. I am a writer, not a visual artist, and to be perfectly honest, if the quality of these books depended on my ability to draw, you would be looking at a stick figure in a firefighter hat right now, and no child deserves that. A series that explores the full breadth of human work and wonder deserves imagery as vivid and varied as the careers it celebrates. What I can promise is that the images are never simply generated and dropped onto the page. Each one is carefully directed, refined, and in nearly every case substantially edited to match the spirit of the words beside it. Real thought and real creative care go into every illustration, and the vision throughout is entirely my own. I have selected illustration tools thoughtfully, prioritizing those that operate within established ethical frameworks for training data and creator compensation. The ethical landscape around AI-assisted creation continues to evolve, and I take seriously the responsibility to engage with it carefully.

THE HEART OF IT

What I hope, more than anything, is that somewhere a child will open one of these books and feel a small door swing open inside them. I hope they will see someone doing work that genuinely fascinates them and quietly think, *maybe me*. I hope a parent or teacher will share a page and find themselves in a conversation that neither of them saw coming. These are books built to accompany a child through the long, beautiful work of wondering, learning, and becoming, and I am grateful you have brought one of them home. Thank you also for the thoughtful feedback and warm support that readers have offered along the way.

— Linda Soules

www.ingramcontent.com/pod-product-compliance
Lightning Source LLC
Chambersburg PA
CBHW041323210926
51847CB00002B/31
* 9 7 8 1 9 7 2 7 6 6 2 1 7 *